FLORA OF TROPICAL EAST AFRICA

———

GRAMMITIDACEAE

B. S. Parris[1]

Usually epiphytic ferns, sometimes terrestrial and lithophytic, occasionally facultative rheophytes. Rhizomes erect to long-creeping, usually scaly. Stipes glabrous or hairy, rarely with scales at the base. Laminae simple to 3-pinnatifid, glabrous or hairy; veins free or rarely with occasional anastomoses, endings sometimes marked by hydathodes on adaxial surface of lamina. Sori without indusia, abaxial, sometimes sunken in pits, pouches or grooves which may be marginal, sometimes protected by the folded pinnae. Sporangia glabrous or with setae, sporangial stalk with one row of cells at base and for most or all of its length. Spores usually globose-tetrahedral and trilete, rarely bilateral and trilete (not in our area), chlorophyllous, sometimes germinating within the sporangium.

A pantropic family of more than 20 genera and over 750 species.

NOTE: Hydathodes, when present, may be as small as 0.1 mm in diameter and some means of magnification, at least a × 10 hand lens and preferably a binocular microscope, is recommended for the accurate identification of most members of the family. **Catenate**: united or linked as in a chain; **clathrate**: latticed or pierced with openings. In general, for all species only the hair types and locations found in all specimens are described. Additional hair types and locations may be found, but do not affect the keys to genera or species.

1. Lamina simple or pinnately divided to a distance 3–4 mm wide
   wide from the midrib 2
   Lamina deeply pinnately divided to a distance 0.1–1.3 mm
   wide from the midrib or pinnate, at least in basal part 3
2. Lamina with translucent to pale red-brown forked hairs
   0.1–0.3(–1.6) mm long or whitish simple eglandular hairs
   0.4–1.5 mm long or glabrous . . . .   1. **Grammitis**
   Lamina with medium to dark red-brown simple eglandular
   hairs 1–3 mm long . . . .   4. **Enterosora**
3. Rhizome scales clathrate, with cell walls medium grey-brown,
   medium to dark brown or dark red-brown, lumen translucent
   to pale brown, sometimes iridescent; hydathodes present on
   vein endings on adaxial surface of lamina 4
   Rhizome scales concolorous, pale to dark red-brown, not
   iridescent; hydathodes absent from vein endings on adaxial
   surface of lamina 5
4. Rhizome scales glabrous or with 1 translucent simple
   glandular hair or 1–3 translucent to pale red-brown simple
   glandular hairs at apex; stipes 1–8 mm long; veins simple;
   sori always 1 per pinna or vein . . . .   2. **Lellingeria**
   Rhizome scales with (1–)2–4(–6) translucent simple glandular
   hairs at apex; stipes over 12 mm long; veins pinnately
   branched; sori usually in 2 rows per pinna, (1–)2–4(–6) per
   row on longest pinnae, rarely only 1 sorus per pinna . . . . . .   3. **Melpomene**

[1] Address of author: Fern Research Foundation, 21 James Kemp Place, Kerikeri, Bay of Islands, New Zealand

1

5. Sori ± midway between pinna mid-vein and margin, without
    white glands when young ....   5. **Zygophlebia**
    Sori much nearer to margin than to pinna mid-vein, covered
    with white glands when young ....   6. **Ceradenia**

1. GRAMMITIS

Sw. in J. Bot. (Schrad.) 1800(2): 3, 17 (1801)

Epiphytic, rarely lithophytic or terrestrial; rhizomes erect to short-creeping, dorsiventral, stipes in two rows (in the Flora area) or radial, with stipes in whorls, not articulated to rhizome nor with phyllopodia (in the Flora area, except in 3. *G. pygmaea*); rhizomes scales that are glabrous or with one to four unbranched hairs of various types at apex. Lamina of fronds simple or pinnately divided, sometimes with dark brown to black sclerotic margin; glabrous or with simple eglandular hairs, pale catenate simple hairs and/or pale catenate forked hairs with pale catenate or medium to dark red-brown simple eglandular branches; veins 1–2-forked (in the Flora area), usually free, rarely anastomosing; hydathodes sometimes absent from vein endings on adaxial surface of lamina. Sori on surface of lamina or very slightly sunken in broad shallow depressions, ± circular to elliptic in outline (in the Flora area), in two rows, one each side of mid-vein (in the Flora area) without receptacular paraphyses; sporangia glabrous (in the Flora area).

A genus of more than 200 species, throughout the montane tropics and southern temperate regions of the Old and New Worlds, currently being revised throughout the Old World. When narrowly construed, a genus of about 22 species, in tropical montane areas of both hemispheres, absent from Asia, Malesia and Australia. In the Flora area only *G. kyimbilensis* belongs to *Grammitis sensu stricto*.

1. Lamina with dark brown to blackish sclerotic margin
    0.1–0.2 mm wide ....   1. *G. kyimbilensis*
    Lamina without dark brown to blackish sclerotic margin 2
2. Rhizomes short-creeping to long-creeping, stipes 1–8 mm
    apart in each row, stipes with whitish simple eglandular
    hairs 0.2–1 mm long; lamina usually with whitish simple
    eglandular hairs 0.4–1 mm on abaxial surface of lamina
    in soral area ....   2. *G. synsora*
    Rhizomes short-creeping, stipes 0.1–1 mm apart in each
    row, glabrous or with translucent to pale red-brown
    simple eglandular hairs ± 0.1 mm long; lamina without
    whitish simple eglandular hairs 3
3. Stipes 1–16 mm long, with translucent to pale red-brown
    simple eglandular hairs ± 0.1 mm long; hydathodes
    present at vein endings on adaxial surface of lamina,
    sometimes obscure; cells in centre of rhizome scales 1–2
    × longer than wide ....   3. *G. pygmaea*
    Stipes winged to base or 1–3 mm long, glabrous;
    hydathodes absent at vein endings on adaxial surface of
    lamina; cells in centre of rhizome scales 3–4 × longer
    than wide ....   4. *G. cryptophlebia*

1. **Grammitis kyimbilensis** (*Brause & Hieron.*) *Copel.* in Philipp. J. Sci. 80(2): 257, fig. 104 (1952); Tardieu, Fl. Madag., Polypod. 2: 74 (1960); Schelpe in Contr. Bolus Herb. 1: 4 (1969); R.J. Johns, Pterid. Trop. E. Afr.: 47–48 (1991); Schippers in Fern Gaz. 14(5): 191 (1993). Type: Tanzania, Rungwe District: Kyimbila, *Stolz* 1028 (B!, lecto. chosen by Parris in K. B. 57, 2: 427 (2002), B!, BM!, K!, M!, WAG!, isolecto.)

Rhizome short-creeping, sometimes branched, stipes 0.2–1.6 mm apart in each row; scales narrowly lanceolate to linear-lanceolate, medium to dark red-brown, glabrous apart from 1–2 pale to dark red-brown simple eglandular hairs at apex, not clathrate, cells in centre of scales 2–4 × longer than wide. Stipe 6–50 mm long with translucent to pale red-brown simple catenate hairs 0.1–0.7 mm long. Lamina narrowly elliptic, linear-elliptic or narrowly lanceolate, 27–155+ mm long, 3–8 mm wide, obtuse to acute or apiculate at apex, cuneate to long-attenuate at base, with dark brown to blackish sclerotic margin 0.1–0.2 mm wide, with translucent to pale red-brown branched catenate hairs 0.1–1.6 mm long on margin; lateral veins 1–(2-) forked, branch endings without hydathodes on adaxial surface of lamina. Sori in middle to apical $\frac{1}{8}$ to ± throughout lamina, 1–31+ in each row. Fig. 1/1–1/4.

TANZANIA. Morogoro District: Uluguru Mts, Lupanga Mt, 14 Nov. 1932, *Schlieben* 2987! & Morogoro–Lupanga track, 16 Aug. 1951, *Greenway & Eggeling* 8616!; Iringa District: Mwanihana Forest Reserve, 10 Oct. 1984, *Thomas* 3817A!
DISTR. **T** 6, 7; Madagascar
HAB. Pendulous epiphyte on tree trunks or on upper branches of tall trees, in montane forest, on granitic rocks, sometimes growing with bryophytes, Hymenophyllaceae and *G. synsora*; 2000–2050 m

SYN. *Polypodium kyimbilense* Brause & Hieron. in E.J. 53: 431–432 (1915)
    P. pseudomarginellum Bonap. in Not. Ptérid. 10: 190 (1920); Christensen in Dansk Bot. Arkiv 7: 156, t. 56, fig. 3–4 (1932). Syntypes: Madagascar, Massif de Manongarivo, *Perrier de la Bâthie* 7471 pro parte (P, syn.) & Ankaziu, 1908, *Perrier de la Bâthie* 7469 (P) & Analamazaotra, *Perrier de la Bâthie* 7525 (P, syn.) & Massif de Manongarivo, *Perrier de la Bâthie* 7705 (BM!, P, US!, syn.)
    Grammitis pseudomarginella (Bonap.) Ching in Bull. Fan Mem. Inst. Biol. 10: 241 (1941)

NOTE. Material cited by R.J. Johns, Pterid. Trop. E. Afr.: 48 (1991) as *G. sp. B aff. kyimbilensis* is referable to *G. cryptophlebia*.

2. **Grammitis synsora** (*Baker*) *Copel.* in Philipp. J. Sci. 80(2): 137 (1952); Tardieu, Fl. Madag., Polypod. 2: 75 (1960); R.J. Johns, Pterid. Trop. E. Afr.: 48 (1991); Schippers in Fern Gaz. 14(5): 192 (1993). Type: Madagascar, Antananarivo, *Gilpin* s.n. (K!, holo.)

Rhizome short-creeping to long-creeping, branched, stipes 1–8 mm apart in each row; scales lanceolate to narrowly lanceolate, dark brown to blackish, glabrous, clathrate, cells in centre of scales 1–2 × longer than wide. Stipes 2–7 mm long, with whitish to pale red-brown simple eglandular hairs (0.2–1 mm long). Lamina narrowly oblanceolate, 18–29 mm long, 3–5 mm wide, obtuse at apex, long-attenuate at base, sometimes glabrous, usually with whitish simple eglandular hairs 0.4–1 mm long on abaxial surface of lamina in soral area; lateral veins simple, each vein-ending marked by a round to slightly elongate hydathode on adaxial surface of lamina, free. Sori in apical $\frac{1}{5}$ to $\frac{1}{2}$ of lamina, 1–8 in each row. Fig. 1/5–1/6.

TANZANIA. Morogoro District: Northern Nguru, Kanga Mountain, 2 Dec. 1987, *Lovett & Thomas* 2807 pro parte! & Nguru Mts, ridge of pass between Mhonde Mission and Maskati Mission, 24 Jan. 1987, *Schippers & De Boer* T 1713! & Nguru Mts, E slopes above Kwamanga Village, 06°10'S 37°35'E, 5 Feb. 1971, *Mabberley* 676A!
DISTR. **T** 6; Madagascar

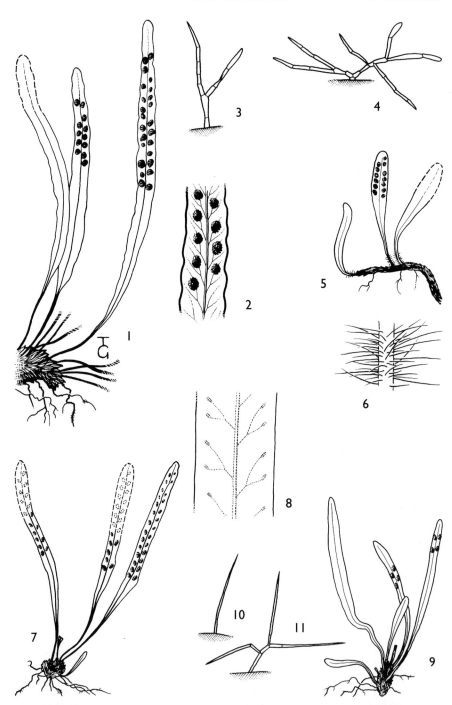

FIG. 1. *GRAMMITIS KYIMBILENSIS* — **1**, habit, × 1; **2**, frond detail, × 2; **3–4**, branched hairs from lamina margin, × 50. *GRAMMITIS SYNSORA* — **5**, habit, × 1; **6**, section of stipe, × 10. *GRAMMITIS PYGMAEA* — **7**, habit, × 1; **8**, lamina adaxial surface with hydathodes. *GRAMMITIS CRYPTOPHLEBIA* — **9**, habit, × 1; **10–11**, simple and branched acicular hairs from lamina margin, × 100. 1–4 from *Stolz* 1028; 5 from *Lovett & Thomas* 2807, 6 from *Schippers & de Boer* T1713; 7–8 from *Schippers* T1625; 9–11 from *Faden* 74/409. Drawn by Tim Galloway.

HAB. High epiphyte, on thin branches of fallen tree, in *Ocotea–Allenblackia–Anthocleista–Podocarpus* forest, sometimes growing with bryophytes, *G. kyimbilensis* and *G. pygmaea*, recorded as rare; 1450–1950 m

SYN. *Polypodium synsorum* Baker in J.L.S. 16: 203–204 (1877); Christensen in Dansk Bot. Arkiv 7: 150–151, t. 56, f. 11, 12 (1932)

NOTE. Specimens from Tanzania have sparser lamina hairs than Madagascan material; sometimes the fronds are glabrous, or the distinctive pale hairs around and within the sori are not visible at maturity.

3. **Grammitis pygmaea** *(Kuhn) Copel.* in Philipp. J. Sci. 80(2): 138 (1952); Iversen in Symb. Bot. Ups. 25, 3: 156 (1991). Type: Réunion [Ins. Borboniae], *Buchinger Herb.* s. n. (B?, holo.)

Rhizome short-creeping, sometimes branched, sometimes reproducing vegetatively by stolons (*Faden* 69/846), stipes 0.1–1 mm apart in each row; scales lanceolate to narrowly lanceolate, pale to medium red-brown or medium yellow-brown, glabrous, not clathrate, cells in centre of scale 1–2 × longer than wide. Stipe 1–16 mm long, with translucent to pale red-brown simple eglandular hairs ± 0.1 mm long. Lamina narrowly oblanceolate to linear-oblanceolate, 14–65 mm long, 2–5 mm wide, bluntly acute to obtuse at apex, long-attenuate at base; glabrous or with translucent to pale red-brown 1–3-forked catenate hairs with simple eglandular branches on both surfaces of lamina and mid-vein and on margin; lateral veins 1-forked, each branch ending marked by a round to elongate hydathode on adaxial surface of lamina, free. Sori in apical $\frac{1}{4}$ to $\frac{2}{3}$ of lamina, 1–19 in each row. Fig. 1/7–1/8.

KENYA. Teita District: Taita Hills, Mraru Ridge, 10 Feb. 1966, *Gillett, Burtt & Osborn* 17166! & Mbololo Hill, Mraru Ridge, 5 July 1969, *Faden et al.* 69/846! & Kasigau, Rukanga Peak, 16 Nov. 1994, *Luke & Luke* 4149
TANZANIA. Lushoto District: W Usambara Mts, Mt Kajuna, hill above Shume Forest Met. Station, 1986, *Schippers* T 1492!; Pare District: S Pare, Mt Shengena, 1 Nov. 1986, *Schippers* T 1625!; Morogoro District: Uluguru Mts, NW slope of Lupanga, 14 Nov. 1970, *Pócs & Harris* 6125/T!
DISTR. **K** 7; **T** 3, 6, 7; Madagascar, Mauritius, Réunion
HAB. Low-level epiphyte, on tree trunks, branches and fallen trees in moist montane forest, sometimes associated with *Dicranolepis, Macaranga, Newtonia*, sometimes growing with bryophytes, lichens and Hymenophyllaceae, recorded as common or very common; 1450–2400 m

SYN. *Polypodium pygmaeum* Kuhn in Filic. Afr.: 152 (1868)
    P. nanodes A.Peter in F.R., Beih. 40: 27, t. 1, f. 7–9 (1929). Type: Tanzania, Lushoto District: W Usambara, Kisimba above Mazumbai, *Peter* 16492 (K!, UC!, iso.)
    Grammitis nanodes (A.Peter) Ching in Bull. Fan. Mem. Inst. Biol. 10: 241 (1941); Schelpe, F.Z., Pterid.: 141, t. 44a (1970) & in Contr. Bolus Herb. 1: 5 (1969); W. Jacobsen, Ferns S Afr.: 296, fig. 212 (1983); J.E. Burrows, S Afr. Ferns & Fern allies: 185, t. 44, fig. 184, 184a, pl. 29, 6 (1990); R.J. Johns, Pterid. Trop. E. Afr.: 48 (1991)

NOTE. The descriptions and illustrations of *G. nanodes* in Schelpe, Contr. Bolus Herb. 1: 5 (1969), Schelpe in F.Z., Pterid.: 141, 142, t. 44 (1970), Jacobsen, Ferns & Fern Allies Southern Africa, 296, fig. 212 (1983) and Burrows, Southern African Ferns & Fern Allies 185, t. 44, fig. 184, 184a, pl. 29, 6 (1990) are referable to *G. cryptophlebia*.

4. **Grammitis cryptophlebia** *(Baker) Copel.* in Philipp. J. Sci. 80(2): 133 (1952); Tardieu in Mém. I.F.A.N. 28: 211, 212, t. 42, fig. 9–10 (1953) & Fl. Madag., Polypod. 2: 76 (1960); R.J. Johns, Pterid. Trop. E. Afr.: 47, 48 (1991). Type: Madagascar, between Tamatave and Antananarivo, *Kitching* s.n. (K!, holo.)

Rhizomes short-creeping, not branched, stipes 0.1–0.2 mm apart in each row; scales narrowly lanceolate to linear-lanceolate, pale to medium red-brown, glabrous, not clathrate, cells in centre of scale 3–4 × longer than wide. Stipes winged to base or 1–3 mm long, glabrous or with translucent to pale red-brown catenate simple acicular hairs and 1–2-forked catenate acicular hairs. Lamina narrowly oblanceolate to linear-oblanceolate, 12–76+ mm long, 2–4 mm wide, bluntly acute to obtuse at apex, long-attenuate at base, with translucent to pale red-brown catenate simple acicular hairs and 1–4-forked translucent to pale red-brown catenate acicular hairs on margin; lateral veins 1-forked, free, branch endings without hydathodes on adaxial surface of lamina. Sori in apical $^1/_8$ to $^3/_4$ of lamina, 1–4 in each row. Fig. 1/9–1/11.

KENYA. Kiambu District: Kikuyu Escarpment Forest, 18 Apr. 1971, *Faden 71/280*!
TANZANIA. Morogoro District: Uluguru Mts, *Goetze* s.n.! & Uluguru Mts, Morningside to Bondwa summit road, 3 April 1974, *Faden 74/399*! & Bondwa summit, 26 Sep. 1970, *Faden 70/648*!
DISTR. **K** 4; **T** 6; São Tomé, Malawi, Mozambique, Zimbabwe, Madagascar
HAB. Epiphyte, on upper branches of tall trees, in moist montane forest, associated with *Cyathea, Podocarpus, Allanblackia uluguruensis, Balthasaria (Melchiora) schliebenii, Cussonia lukwangulensis, Lobelia lukwangulensis, Ocotea usambarensis, Polyscias stuhlmannii, Syzygium micklethwaithii*, sometimes growing with bryophytes; 1900–2150 m

SYN. *Polypodium cryptophlebium* Baker in J.B. 18: 370 (1880)
     P. rutenbergii Luerssen in Abhandl. Bot. Ver. Bremen 7: 48, pl. 1 fig. 1–2 (1882); Christensen in Dansk Bot. Arkiv 7: 150, t. 56, fig. 21, 22 (1932). Type: Madagascar, Ambatondrazaka, *Rutenberg* in herb. *Luerssen* 9672 (BREM?, L?, P?, drawing in B!)
     P. molleri Baker in Bol. Soc. Brot. 4: 154, t. 2, f. b (1887). Type: São Tomé, Pico, *Moller* in Herb. Hort Bot. Conimbrensis no. 51 (BM!, BR!, K!, UC, iso.)
     Grammitis molleri (Baker) Schelpe in Bol. Soc. Brot. Ser. 2a, 40: 162 (1966)
     [*G. pygmaea*; sensu Schippers in Fern Gaz. 14(5): 192 (1993), *non* (Mett.) Copel. pro parte]
     G. sp. A. Faden in U.K.W.F. ed. 2: 23 (1994)

NOTE. The hairs described here as catenate acicular actually consist of a rather short basal cell and a much longer acicular apical cell. They are not the typical catenate simple eglandular hairs of other species because they are only two cells long and the cells are unequal in length. Similar hairs occur in the group of *G. billardierei* Willd., with which this species seems allied, indeed the low number of indurated annulus cells is similar to that of *G. stenophylla* Parris.

2. **LELLINGERIA**

A.R.Sm. & R.C.Moran in Amer. Fern J. 81(3): 76 (1991)

Epiphytic or lithophytic; rhizomes usually erect to short-creeping, rarely long-creeping, dorsiventral (described as radial in the original generic description, but the type species, *L. apiculata* (Klotzsch) A.R.Sm. & R.C.Moran, clearly has a dorsiventral rhizome). Stipes in two rows, not articulated to rhizome, phyllopodia absent; rhizomes with dark brown to dark red-brown lanceolate to narrowly lanceolate strongly clathrate glabrous (or with 1–3 hairs only at apex) or ciliate (not in Flora area) scales. Lamina of fronds shallowly to deeply pinnately divided to 1-pinnate-pinnatifid (not in Flora area), the fertile apical portion sometimes entire or less divided than the sterile part; nearly always with 1(–2)-forked catenate hairs with simple eglandular branches and glandular apex, hairs translucent to pale red-brown, usually ± 0.1–0.2 mm long (in the Flora area); veins simple or to pinnately branched (not in Flora area), branchlets simple, free; hydathodes present on vein endings on adaxial surface of lamina. Sori on surface of lamina or sunken in depressions, ± circular to broadly elliptic in outline, sometimes coenosoroid, in two rows, one each side of mid-vein (in the Flora area) rarely with branched hairs as on the lamina present as receptacular paraphyses; sporangia glabrous or with simple eglandular hairs scattered adjacent to annulus in apical part (not in Flora area).

A genus of ± 60 species, occurring mainly in tropical montane areas of the New World.

1. Fertile and sterile parts of lamina similarly divided . . . . .   1. *L. oosora*
   Apical fertile part of lamina less deeply divided than basal
      sterile part 2
2. Frond including stipe without forked hairs (**T** 6, Nguru
      Mts, endemic) . . . .   2. *L. rupestris*
   Frond including stipe with forked hairs 3
3. Apical $^2/_3$ of lamina ± undivided, pinnae 1–3 pairs, longest
      pinnae 1 mm long, 1–2 mm wide . . . .   3. *L. paucipinnata*
   Apical $^1/_5$ to $^1/_2$ of lamina ± lobed, pinnae 6–18 pairs,
      longest pinnae (1–)2–3 × 1 mm . . . .   4. *L. strangeana*

1. **Lellingeria oosora** (*Baker*) *A.R.Sm. & R.C.Moran* in Amer. Fern J. 81(3): 85 (1991). Type: São Tomé, Pico, *A.Moller* 46 (K!, lecto. chosen by Parris in K.B. 57, 2: 428 (2002); B!, BM!, UC, isolecto.)

Rhizome short-creeping, stipes 0.1–0.2 mm apart in each row. Stipe 2–6 mm long, with 1-forked catenate hairs and simple glandular hairs. Lamina linear in outline, 49–110 mm long, 4–7 mm wide, obtuse to acute at apex, cuneate to long-attenuate at base, pinnate, pinnae 22–49 pairs, longest pinnae 2–4 mm wide and 1–2 mm long, oblong, bluntly acute at apex, with 1-forked catenate hairs on both surfaces of rachis, and simple glandular hairs on both surfaces of rachis and on adaxial surface of lamina; lateral veins simple, each branch ending marked by an elongate hydathode on adaxial surface of lamina, free. Sori on 13–33 pairs of pinnae in apical $^1/_2$ to $^4/_5$ of lamina, 1 per pinna, in basal to middle $^1/_2$ to $^3/_4$ of pinna. Fig. 2/1–2/2.

TANZANIA. Rungwe District: Kyimbila, Lukeke, 13 Dec. 1911, *Stolz* 1027!
DISTR. **T** 7; Sierra Leone, Guinea, Liberia, Ivory Coast, Cameroon, Bioko, São Tomé, Gabon, Rwanda, Malawi, Madagascar
HAB. Epiphyte in montane forest, growing with bryophytes; no altitude given

SYN. *Polypodium oosorum* Baker in Henriq., Bol. Soc. Brot. 4: 154 (1886); Christensen in Dansk Bot. Arkiv 7: 153, t. 57, fig. 6, 8 (1932); Tardieu in Mém. I.F.A.N. 28: 212, t. 42, fig. 102, 223 (1953)
   Xiphopteris oosora (Baker) Alston in Bol. Soc. Brot. sér. 2a, 30: 26 (1956) & Ferns W.T.A., ed. 2 Suppl.: 45 (1959); Tardieu, Fl. Madag., Polypod. 2: 82 (1960); Schelpe in F.Z., Pterid.: 143 (1970); Tardieu, Fl. Cameroun 3: 325 (1964); Schelpe in Contr. Bolus Herb. 1: 12 (1969) pro parte; Benl in Acta Bot. Barcinon. 33: 21 (1982); Pic.Serm. in B.J.B.B. 53 (1/2): 199 (1983); R.J. Johns, Pterid. Trop. E. Afr.: 49 (1991); Kornaś et al. in Zesz. Nauk. Uniw. Jagiellon. Prace Bot. 25: 43 (1993); Schippers in Fern Gaz. 14(5): 192 (1993); Cable & Cheek, Pl. Mt Cameroon: 190 (1998)
   Grammitis oosora (Baker) J.E. Burrows, S Afr. Ferns & Fern Allies: 186 (1990)

NOTE. Schelpe (in Contrib. Bolus Herb. 1: 12, 1969) cites the following collections from Mlanje mountain, Malawi, as *X. oosora*, but they are referable to *Lellingeria hildebrandtii* (Hieron.) A.R.Sm. & R.C.Moran: *Brass* 16670, *Newton & Whitmore* 402, *Richards* 16643. He also cites *Brass* 16670 as *X. oosora* in Schelpe in F.Z., Pterid.: 144, 1970. Material illustrated by Burrows (S. Afr. Ferns & Fern Allies: 185, fig. 44 & pl. 30, 2, 1990) as *G. oosora* also belongs to *L. hildebrandtii* and his report of *G. oosora* from Zimbabwe is apparently based upon material that I have identified as *Melpomene flabelliformis*.

2. **Lellingeria rupestris** *Parris* in K.B. 57, 2: 429 (2002). Type: Tanzania, Morogoro District, Nguru Mountains, near Maskati [Muskat] mission, *Schippers* T 1703 (K!, holo., WAG!, iso.)

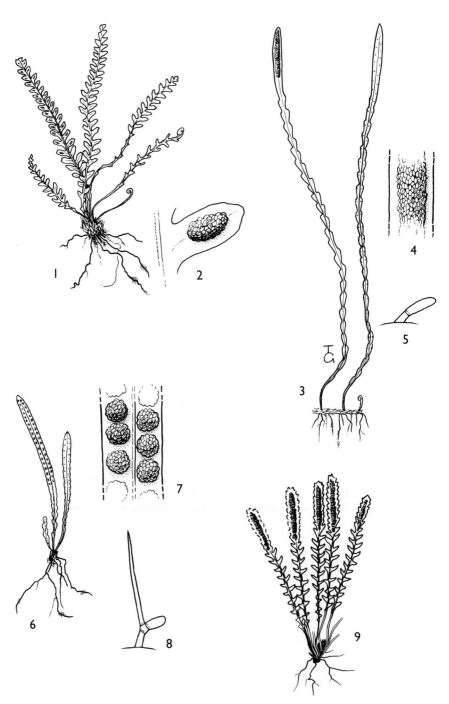

FIG. 2. *LELLINGERIA OOSORA* — **1**, habit, × 1; **2**, sorus on pinna, × 7. *LELLINGERIA RUPESTRIS*
— **3**, habit, × 1; **4**, sorus, × 7; **5**, 2-celled hair, × 100. *LELLINGERIA PAUCIPINNATA* — **6**,
habit, × 1; **7**, sori, × 7; **8**, branched hair, × 100. *LELLINGERIA STRANGEANA* — **9**, habit, ×
1. 1–2 from *Stolz* 1027; 3–5 from *Schippers* T1703; 6–8 from *Hemp* 1636; 9 from *Thomas* 3818.
Drawn by Tim Galloway.

Rhizome long-creeping, stipes 2–5 mm apart in each row. Stipes 6–17 mm long, glabrous or with simple glandular hairs. Lamina linear in outline, 62–110+ mm long, 1–2 mm wide, bluntly acute at apex, long-attenuate at base, deeply pinnately divided to a distance 0.1–0.2 mm wide from the midrib in sterile part of lamina, apical fertile $\frac{1}{8}$–$\frac{1}{10}$ of lamina entire, pinnae 19–37 pairs, longest pinnae ± 1 mm long, 1 mm wide, triangular, bluntly acute to obtuse at apex, with simple glandular hairs on abaxial surface of rachis; lateral veins simple, free, each vein ending marked by a round to elongate hydathode on adaxial surface of lamina. Sori coenosoroid between veins and across rachis at maturity, on 2–8 pairs of veins, in apical $\frac{1}{8}$–$\frac{1}{10}$ of lamina, 1 per frond. Fig. 2/3–2/5.

TANZANIA. Morogoro District: Nguru Mountains, between Mhonda & Maskati [Muskat] missions, 24 Jan. 1987, *Schippers* T 1703!
DISTR. **T** 6; only known from the type locality
HAB. On rocks in fast flowing river, growing with bryophytes, *Hymenophyllum tunbrigense* and *H. peltatum*; ± 1900 m

SYN. [*Xiphopteris strangeana* sensu R.J. Johns, Pterid. Trop. E. Afr., 49 (1991), *non* Pic.Serm. in Webbia 27: 453, f. 23 (1973)]
    [*X.* cf. *hildebrandtii* sensu Schippers in Fern Gaz. 14(5): 192 (1993), *non* Tardieu in Hedwigia 44: 91 (1905)]

NOTE. *Lellingeria rupestris* resembles *L. strangeana* in having the apical fertile part of the lamina undivided, but differs in having only simple glandular hairs rather than 1–2-forked catenate hairs with simple eglandular branches and glandular apex.
    Schippers (Fern Gaz. 14(5): 192, 1993) records this collection as *X.* cf. *hildebrandtii* (Hieron.) Tardieu and notes "Known only from a fast flowing river in Nguru at 1910 m (*Schippers & de Boer* 1703) where it was locally common. A very interesting species growing on stones and small rocks just above the water level. All fronds face towards the water and plants thereby grow in a circle around the rock's edge. This species is not known from other places on the African continent. Our plants resemble the Madagascan *X. hildebrandtii* in rhizome scales, frond hairs and hydathodes but differ in the broad relatively undivided lamina (pers. comm. Dr Barbara Parris)."

3. **Lellingeria paucipinnata** *Parris* in K.B. 57, 2: 428 (2002). Type: Tanzania, Moshi District, Kilimanjaro, Kikafu River Gorge, *Hemp* 1636 (K!, holo.)

Rhizome short-creeping, stipes ± 0.1 mm apart in each row. Stipe 5–6 mm long with 1-forked catenate hairs. Lamina linear-oblanceolate to linear-elliptic in outline, ± 40 mm long, 2–3 mm wide, bluntly acute to obtuse at apex, long-attenuate at base, pinnate at base or basal $\frac{1}{3}$ divided to a distance 0.1–0.2 mm from the midrib, apical $\frac{2}{3}$ ± entire, pinnae 1–3 pairs and 3–5 pairs of apical lobes, , longest pinnae ± 1 mm long, 1–2 mm wide, broadly triangular, bluntly acute to obtuse, with catenate 1–2-forked hairs with simple eglandular branches and glandular apex sparse to scattered on both surfaces of lamina and rachis; lateral veins simple, free, each vein ending marked by an elongate hydathode on adaxial surface of lamina. Sori in ± entire apical $\frac{2}{3}$ of lamina, not on pinnae, 18 in each row. Fig. 2/6–2/8.

TANZANIA. Moshi District: Kilimanjaro, Kikafu River Gorge, 3 March 1997, *Hemp* 1636!
DISTR. **T** 2; not known elsewhere. See Note below.
HAB. Epiphyte in ericaceous zone, or on boulder in stream bed, growing with bryophytes; 2310–2550 m

SYN. *Xiphopteris* sp. A, Johns, Pterid. Trop. E. Afr.: 49 (1991) — see Note
    X. sp. B, Faden in U.K.W.F. ed. 2: 23 (1994) — see Note

NOTE. *Lellingeria paucipinnata* is possibly known only from the type collection, although two collections cited as *X. sp. B* by Faden in U.K.W.F. ed. 2, 23 (1994) and as *X.* sp. A by Johns, Pterid. Trop. E. Afr.: 49 (1991) may belong here. They are *Faden et al.* 69/524 and *Faden* 70/97, neither of which I have seen. Hemp, on the label on the type collection, calls it *Xiphopteris* spec. A. Johns 1991 (= *X.* spec. B. Faden 1994 [sic]).

See under *L. strangeana* for differences between this and *L. paucipinnata.*

4. **Lellingeria strangeana** (*Pic.Serm.*) *A.R.Sm. & R.C.Moran* in Amer. Fern J. 81(3): 87 (1991). Type: Kenya, Teita District, Mt Kasigau, upper slopes of summit peak, *Faden et al* 71/186 (FT!, holo., BOL!, BR!, EA, K!, MO!, US!, iso.)

Rhizome ± erect to short-creeping, stipes 0.1–0.3 mm apart in each row. Stipe 1–7 mm long, with 1–2-forked catenate hairs. Lamina linear, linear-elliptic, linear-oblanceolate or narrowly oblanceolate in outline, 17–76 mm long, 2–5 mm wide, acute to obtuse at apex, long-attenuate at base, pinnate or deeply pinnately divided to a distance 0.1–0.2 mm from the midrib in basal part of lamina, lobed ± $^1/_3$ to less than halfway to rachis in apical part, ± lobed portion in apical $^1/_5$ to $^1/_2$ of lamina, pinnae 6–18 pairs and 2–20 pairs of apical lobes, longest pinnae broadly to narrowly triangular, narrowly oblong, oblong or oblanceolate-oblong, 1–3 mm long, 1 mm wide, obtuse to acute at apex, with 1–2-forked catenate hairs on abaxial surface of rachis; lateral veins simple, free, each vein ending marked by an elongate hydathode on adaxial surface of lamina. Sori in apical $^1/_4$ to $^1/_2$ of lamina, on 0–4 pairs of pinnae and 2–20 pairs of apical lobes, 1 per pinna or apical lobe, on basal $^1/_3$ to $^1/_2$ of pinna to ± throughout. Fig. 2/9.

KENYA. Kericho District: Kericho, African Highlands Co. Dam, 25 Sept 1966, *Strange* 220!; Teita District: Mbololo, Mraru Ridge, 5 July 1969, *Faden et al.* 69/850! & Mt Kasigau, Rukanga to summit, 6 Feb. 1971, *Faden et al.* 71/186!
TANZANIA. Pare District: Kwizu Forest summit, 13 Sep. 1986, *Schippers* T1562!; Lushoto District: W Usambara Mts, Mt Kajuna, hill above Shume Forest meteorological station, 1986, *Schippers* T1495!; Iringa District: Udzungwa Mts, Mwanihana Forest Reserve, 10 Oct. 1984, *Thomas* 3847A!
DISTR. **K** 5, 7; **T** 3, 6, 7; not known elsewhere
HAB. Epiphyte on tree trunks and branches, and on fallen logs, in montane forest, sometimes with *Syzygium micklethwaithii* and *Rapanea* dominant but dwarfed, associated with *Newtonia, Macaranga* and *Dicranolepis*, growing with bryophytes, recorded as occasional, common and abundant; 1600–2400 m

SYN. [*Xiphopteris myosuroides* sensu Schelpe in Contr. Bolus Herb. 1: 10 (1969), *non* Sw.]
    X. strangeana Pic.Serm. in Webbia 27: 453, f. 23 (1973); Iversen in Symb. Bot. Ups. 29, 3: 110, 156 (1991); R.J. Johns, Pterid. Trop. E. Afr.: 49 (1991) pro parte; Schippers in Fern Gaz. 14(5): 192 (1993); Faden in U.K.W.F. ed. 2: 23 (1994)

NOTE. Material recorded as *X. myosuroides* by Johns (Pterid. Trop. E. Afr.: 48 (1991): Tanzania, T 3, T 6, *Pócs* 6827) is probably referable to this species.
    Johns (Pterid. Trop. E. Afr.: 49 (1991)) refers *Schippers* T1703 to *X. strangeana*, but it belongs to *L. rupestris*.
    Lellingeria strangeana is related to *L. paucipinnata*, from which it differs in having more and longer pinnae, 6–18 pairs, the longest (1–)2–3 mm long and 1 mm wide in the former, 1–3 pairs, the longest 1 mm long, 1–2 mm wide in the latter, and a proportionately shorter less divided apical portion, $^1/_5$ to $^1/_2$ of lamina length and lobed ± $^1/_3$ to less than $^1/_2$ way to the rachis in the former, $^2/_3$ of lamina length and ± entire in the latter.

3. MELPOMENE

A.R.Sm. & R.C.Moran in Novon 2(4): 426 (1992)

Epiphytic, terrestrial or lithophytic; rhizomes erect, short-creeping to long-creeping, dorsiventral, stipes in two rows, not articulated to rhizome; rhizomes with strongly clathrate scales that are glabrous apart from one or more glandular hairs at the apex, scales sometimes extending to the basal part of the stipe. Fronds pinnate, pinnatifid or pinnatisect; with pale catenate branched hairs, the cross-walls evident, often with glandular apex to branches, and red-brown simple eglandular hairs; veins

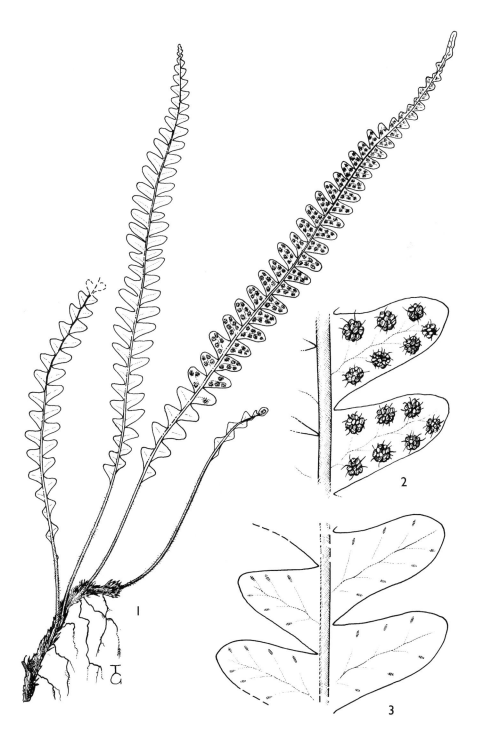

FIG. 3. *MELPOMENE FLABELLIFORMIS* — **1**, habit, × 1; **2**, pinnae, abaxial, showing sori, × 5; **3**, pinnae, adaxial, showing hydathodes, × 5. 1–3 from *Hedberg* 2025. Drawn by Tim Galloway.

pinnately branched in pinnae, branchlets simple, free; ± circular to elongate hydathodes present on vein endings on adaxial surface of lamina. Sori on surface of lamina or slightly sunken in broad shallow depressions, ± circular to broadly elliptic in outline, usually in 2 rows per pinna, lacking simple or branched glandular receptacular paraphyses, but sometimes with simple eglandular receptacular paraphyses; sporangia glabrous.

A genus of about 20 species, mainly in the tropical montane areas of the New World.

Melpomene flabelliformis (*Poir.*) *A.R.Sm. & R.C.Moran* in Novon 2(4): 430 (1992). Type: Réunion [Bourbon], *Commerson in Jussieu* 1098-C (P, lecto, chosen by Morton in Contr. U.S. Nat. Herb. 38: 57 (1967); US, photo.)

Rhizome long-creeping, sometimes branched. Stipes 1.2–10.2 cm long, with 1–4-forked catenate hairs 0.1–0.7 mm long. Lamina linear-elliptic or narrowly elliptic in outline, 2.3–33+ cm long, 0.5–2.2 cm wide, acute to obtuse at apex, truncate to long-attenuate to very narrow stipe wing at base, pinnate or very deeply pinnately divided to a distance 0.1–0.3 mm from the midrib, pinnae 9–56+ pairs, longest pinnae triangular to oblong, 3–10 mm long, 2–5 mm wide, obtuse to bluntly acute at apex; with 1–7-forked catenate hairs 0.1–0.7 mm long on abaxial surfaces of rachis, nearly always with simple eglandular hairs 0.3–2.5 mm long on abaxial surface of lamina especially amongst sori. Sori on 1–51 pairs of pinnae, in apical $^1/_4$ of lamina to ± through except basal pair of pinnae, in 2 rows per pinna, 1 each side of pinna midvein, or solitary on pinnae, 1–6 in each row on longest pinnae, in apical or basal $^1/_2$ to $^3/_4$ of pinnae to ± throughout. Fig. 3.

UGANDA. Toro District: Ruwenzori, Mobuku Valley, Oct. 1908, *Kassner* 3138! & Ruwenzori, idem, on ridge above Nyabitaba hut, 15 Jan. 1967, *Smith* 4596!; Mbale District: Mt Elgon, Nametata R. basin, 6 Nov. 1916, *Snowden* 494!
KENYA. Nyandarua/Aberdare Mts, 31 Jan. 1922, *Fries* 1325!; Meru District: E slope of Mt Kenya, Kirui cone, 13 April 1969, *Faden et al* 69/518!; South Nyeri District: Mt Kenya, South Kamweti Track above Kamweti Forest Station, 9 Jan. 1972, *Archer in EA* 15046!
TANZANIA. Moshi District: Kilimanjaro, Marangu, 2 Feb. 1893, *Volkens* 1210! & Kilimanjaro, near Kibosho, 1906, *Daubenberger in Rosenstock's Filic. Afr. or. germ. exsicc.* 27!; Morogoro District: Uluguru Mts, S Uluguru Forest reserve, edge of Lukwangule Plateau, 17 March 1953, *Drummond & Hemsley* 1669!
DISTR. **U** 2, 3; **K** 3, 4; **T** 2–4, 6, 7; Cameroon, Bioko, Congo (Kinshasa), Rwanda, Ethiopia, Malawi, Mozambique, Zimbabwe, South Africa, Madagascar, Mauritius, Réunion
HAB. Usually a low epiphyte, often in moss cushions on tree trunks and branches, in giant heath or *Podocarpus/Hagenia* forest to *Hagenia* woodland, afroalpine bush, bamboo zone, sometimes on rocks of old lava flows; 1000–4200 m

SYN. *Polypodium flabelliforme* Poir. in Lam., Encycl. Meth. Bot. 5: 519 (1804)
    P. rigescens Willd., Sp. Pl. ed. 4, 5: 183 (1810); Christensen in Dansk Bot. Arkiv 7: 154 (1932). Type: Réunion [Bourbon] (B-W, holo., photo!, P-JU, iso.)
    Ctenopteris rigescens (Willd.) J. Sm., Hist. Fil., 184 (1875); Tardieu, Fl. Madag., Polypod. 2: 84, t. 20, fig. 1–2 (1960); Tardieu, Fl. Cameroun 3: 327, t. 52, fig. 1–3 (1964); Pic.Serm. in B.J.B.B. 53(1/2): 197 (1983)
    Xiphopteris rigescens (Willd.) Alston in Bol. Soc. Brot. ser. 2a, 30: 27 (1956) [err. '*nigrescens*']; Alston, Ferns W.T.A., ed. 2 Suppl.: 45 (1959); R.J. Johns, Pterid. Trop. E. Afr.: 48, 49 (1991)
    X. flabelliformis (Poir.) Schelpe in Bol. Soc. Brot. ser. 2a, 41: 217 (1967) & in Contrib. Bolus Herb. 1: 10 (1969) & F.Z., Pterid.: 142, t. 44b (1970); Benl in Acta Bot. Barcinon. 33: 22 (1982); W. Jacobsen, Ferns Sthn Afr.: 298, fig. 214 (1983); Schelpe, F.S.A. Pteridophyta: 150, fig. 48, 3, 3a, 152 (1986); Iversen in Symb. Bot. Upsal. 29, 3: 156 (1991); R.J. Johns, Pterid. Trop. E. Afr.: 48, 49 (1991); Kornaś et al. in Zesz. Nank. Uniw. Jagiellon. Prace Bot. 25: 43 (1993); Schippers in Fern Gaz. 14(5): 192 (1993); Faden in U.K.W.F. ed. 2: 23 (1994); Cable & Cheek, Pl. Mt Cameroon: 190 (1998)
    Grammitis flabelliformis (Poir.) C.V.Morton in Contr. U.S. Nat. Herb. 38: 57 (1967)
    G. rigescens (Willd.) Lellinger in Proc. Biol. Soc. Wash. 98(2): 383 (1985); J.E. Burrows, S Afr. Ferns & Fern allies: 185, fig. 44, 186, t. 30, 1 (1990)

NOTE. Wood notes (on *Wood* 267) that plants usually form single tufts alone, while Greenway notes (on *Greenway* 3174) that this species forms scattered clumps up to 3–4 feet square.

Many dried specimens have stained their mounting paper yellow; the stained paper is faintly aromatic.

Melpomene flabelliformis is variable with respect to spore and sporangium size, and presence and distribution of hairs, but the variations have no geographical basis.

4. **ENTEROSORA**

Baker in Timehri 5: 218 (1886); Bishop & Smith in Syst. Bot. 17(3): 347 (1992)

Epiphytic or lithophytic; rhizomes short-creeping, dorsiventral or radial, stipes in two rows, articulated to rhizome, or in whorls of 3, not articulated to rhizome; rhizomes with brown non-clathrate scales that may be ciliate or glandular. Fronds simple to broadly lobed to $\frac{1}{5}$ the way to the mid-vein, with red-brown simple eglandular hairs; veins 1–3-forked or pinnately branched, sometimes anastomosing; hydathodes absent from vein endings on adaxial surface of lamina. Sori on surface of lamina, slightly sunken in broad shallow depressions, or deeply sunken in steep-sided depressions (not in the Flora area), ± circular to broadly elliptic (to elongate, not in the Flora area), sometimes with simple or branched glandular receptacular paraphyses that are not evident at soral maturity, rarely with simple eglandular receptacular paraphyses; sporangia glabrous.

A genus of 10 species, mainly in the tropical montane areas of the New World.

Rhizomes dorsiventral; stipes in 2 rows, articulated to rhizome,
   > 10 mm long . . . .   1. *E. barbatula*
Rhizomes radial; stipes in whorls of 3, not articulated to rhizome,
   < 10 mm long . . . .   2. *E. sprucei*

1. **Enterosora barbatula** (*Baker*) *Parris* in K.B. 57, 2: 426 (2002). Type: Réunion [Bourbon], *Balfour* s.n. (K!, lecto., chosen by Schelpe, 1969)

Rhizome short-creeping, sometimes branched, dorsiventral, stipes in 2 rows, 0.5–2 mm apart in each row, articulated to rhizome, phyllopodia 0.4–0.9 mm high; scales narrowly lanceolate, with simple glandular and catenate simple glandular hairs ± 0.1 mm long on margin. Stipe 32–108 mm long, with simple eglandular hairs 1.2–2.5 mm long. Lamina narrowly elliptic to linear-elliptic in outline, 73–320+ mm long, 7–18 mm wide, obtuse to bluntly acute at apex, cuneate to long-attenuate at base, smaller fronds entire, larger fronds pinnately divided to 3–4 mm from mid-vein, up to $\frac{1}{2}$ of lamina width, lobes up to 10 mm long; texture spongiose-coriaceous to coriaceous; with simple eglandular hairs 1–2.1 mm long on both surfaces of lamina and on margin; lateral veins 1–3-forked or pinnately branched in lobes, free or anastomosing, especially near margin in larger fronds, sometimes along mid-vein. Sori in 2–6 rows, 1–3 on each side of mid-vein, in apical $\frac{1}{5}$ to ± throughout lamina, 14–106 in first row (nearest mid-vein), 2–20 in second row, 1–7 in third row. Fig. 4/3–4/5.

TANZANIA. Pare District: S Pare Mts, near Mkongo R., *de Boer* 141; Morogoro District: Parata Pass, 18 Oct. 1932, *Schlieben* 2827! & Uluguru North Forest Reserve, main ridge of North Uluguru Mountain close to Lupanga Peak, 9 Dec. 1980, *Hall* s.n.!
DISTR. **T** 3, 6; Madagascar, Mauritius and Réunion
HAB. Epiphyte on tree trunks, sometimes under deeply shaded leaning tree bole, or on overhanging wet rocks, in montane forest on Basement Complex rocks (granulites), in areas with 3000 mm rainfall per annum, sometimes growing with bryophytes and Hymenophyllaceae; 1900–2100 m

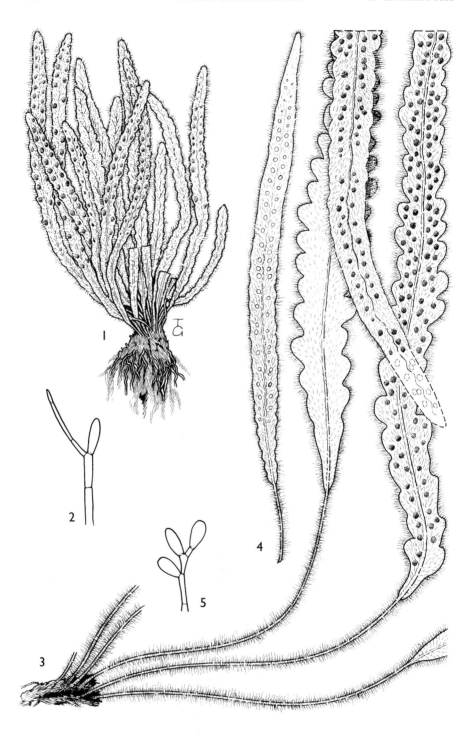

FIG. 4. *ENTEROSORA SPRUCEI* — **1**, habit, with lobed fronds, × 1; **2**, branched receptacular paraphysis, × 80. *ENTEROSORA BARBATULA* — **3**, habit, × 4; **4**, entire frond, × 1; **5**, branched receptacular paraphysis, × 80. 1–2 from *Schlieben* 2827b; 3–5 from *Schlieben* 2827. Drawn by Tim Galloway.

SYN. *Polypodium barbatulum* Baker in Hook. & Baker, Syn. Fil., 323 (1867); Christensen in Dansk
Bot. Arkiv 7: 152, t. 57, fig. 20–22 (1932)
    P. poolii Baker in J.L.S. 15: 419 (1876). Type: Madagascar, Antananarivo, *Pool* s.n. (K!, holo.)
    P. uluguruense Reimers in N.B.G.B. 11: 932 (1933). Type: Tanzania, Morogoro District,
Uluguru Mts, Parata Pass, *Schlieben* 2827 (B!, lecto., chosen by Parris in 2002, B!, BM!,
BR!, M!, isolecto.)
    Grammitis barbatula (Baker) Ching in Bull. Fan Mem. Inst. Biol. 10: 240 (1941); Copeland
in Philipp. J. Sci. 80(2): 135 (1952); Tardieu, Fl. Madag., Polypod. 2: 78 (1960); Schelpe
in Contr. Bolus Herb. 1: 5 (1969); R.J. Johns, Pterid. Trop. E. Afr.: 47 (1991); Schippers
in Fern Gaz. 14(5): 191 (1993)
    G. poolii (Baker) Copel. in Gen. Fil. 211 (1947)
    G. uluguruensis (Reimers) Copel. in Philipp. J. Sci. 80(2): 135 (1952)

NOTE. Tanzanian material is generally larger than that from Madagascar, Mauritius and
Réunion. The hairs on the margin of the rhizome scales are usually also present on the
abaxial surface and are paler and longer in Tanzanian specimens than in those from
elsewhere. These distinctions seem insufficient to maintain *P. ulugurense* as separate from *E
barbatula* at any rank.

2. **Enterosora sprucei** (*Hook.*) Parris in K.B. 57, 2: 426 (2002). Type: Peru, near
Tarapoto, *Spruce* 4746 (K!, lecto., chosen by Parris in 2002)

Rhizome ± erect, not branched, radial, stipes in whorls of 3, not articulated to
rhizome, phyllopodia absent; scales broadly ovate to narrowly lanceolate, with simple
eglandular hairs 0.1–0.3 mm long solitary at apex and sometimes scattered on
margin. Stipe 3–7 mm long, with simple eglandular hairs 1.7–2.8 mm long and
1–2-forked translucent to pale red-brown catenate hairs 0.1–0.3 mm long.. Lamina
narrowly elliptic to linear-elliptic, 37–101 mm long, 3–7 mm wide, obtuse to bluntly
acute at apex, long-attenuate at base, entire or crenulate, lobes 0.1–4 mm long,
divided up to $^2/_3$ of distance to mid-vein; with simple eglandular hairs 1.2–3 mm long
on both surfaces of lamina and on margin, translucent 1–2-forked catenate hairs
0.1–0.7 mm long on abaxial surface, and translucent catenate simple hairs ± 0.1 mm
long on abaxial surface; lateral veins simple or 1-forked, rarely pinnately divided in
lobes, free. Sori usually in 2 rows, 1 each side of mid-vein, rarely up to 4 on each lobe,
in apical $^1/_5$ to $^4/_5$ of lamina, 1–17 in each row. Fig. 4/1–4/2.

TANZANIA. Morogoro District: NW Uluguru Mts, Parata Pass, 18 Oct. 1932, *Schlieben* 2827a! &
Uluguru North Forest Reserve, main ridge of North Uluguru Mountain close to Lupanga
Peak, 9 Dec. 1980, *Hall* s.n.!
DISTR. **T** 6; Madagascar; Mexico, Costa Rica, Venezuela, Peru
HAB. Epiphyte on trees, in montane forest on Basement Complex rocks (granulites), in areas
with 3000 mm rainfall per annum; 1900–2100 m

SYN. *Polypodium sprucei* Hook., 2nd Century ferns: t. 10 (1860)
    P. gilpinae Baker in J.L.S. 16: 204 (1877). Type: Madagascar, Antananarivo, *Gilpin* s. n.
(K!, holo.)
    P. microphyllum Baker in K.B. 1897: 299 (1897); Christensen in Dansk Bot. Arkiv 7: 151, t.
56, fig. 7–8 (1932). Type: Madagascar, Ambohimitombo, *Forsyth-Major* 477 (K!, holo.; B!,
BM!, iso.).
    P. poolii sensu Christensen in Dansk Bot. Arkiv 7: 151, t. 56, fig. 9–10 (1932), *non* Baker
    P. pseudopoolii Reimers in N.B.G.B. 11: 934 (1933). Type: Tanzania, Morogoro District,
Uluguru Mts, Parata Pass, *Schlieben* 2827a (B!, holo.)
    Grammitis poolii sensu Copel. in Philipp. J. Bot. 80(2): 134, fig. 12 (1952); Schelpe in Contr.
Bolus Herb. 1: 4 (1969); R.J. Johns, Pterid. Trop. E. Afr.: 48 (1991); Schippers in Fern
Gaz. 14(5): 192 (1993)
    G. gilpinae (Baker) Tardieu, Fl. Madag. Pterid. 5(2): 79 (1960)
    Enterosora gilpinae (Baker) L.E.Bishop & A.R.Sm. in Syst. Bot. 17(3): 359, fig. 3/c–e (1992)

5. ZYGOPHLEBIA

L.E.Bishop in Amer. Fern J. 79(3): 103 (1989)

Usually epiphytic, rarely terrestrial or lithophytic; rhizomes short-creeping, dorsiventral, stipes in two rows, articulated to rhizome, phyllopodia present; rhizomes with brown non-clathrate scales that are usually glandular at least on margin. Fronds pinnate or deeply pinnately divided, with red-brown simple eglandular hairs, catenate branched hairs often with glandular apex to branches, and simple glandular and catenate simple glandular hairs; veins pinnately branched in pinnae, branchlets often forked, sometimes anastomosing; hydathodes absent on vein endings on adaxial surface of lamina. Sori on surface of lamina or slightly sunken in broad shallow depressions, ± circular to broadly elliptic in outline, in two rows per pinna, one each side of pinna mid-vein, with simple or branched glandular receptacular paraphyses, the glands never white when young; sporangia glabrous.

A genus of about 14 species, occurring in tropical montane areas of the New World and Africa including Madagascar.

Rhizome scales 1–1.7 mm+ wide, cells of rhizome scales usually not
   turgid, rarely sub-turgid .... 1. *Z. major*
Rhizome scales 0.3–0.5 mm wide, cells of rhizome scales subturgid
   to turgid .... 2. *Z. devoluta*

1. **Zygophlebia major** (*Reimers*) *Parris* in K.B. 57, 2: 433 (2002). Type: Tanzania, Morogoro District, Uluguru Mts, Bondwa [Bondua], *Schlieben* 3018 (B!, holo.)

Rhizome short-creeping, not branched. Stipes 0.4–2.5 mm apart in each row; scales narrowly lanceolate, 3.8–5.2+ mm long, 1–1.7 mm wide. Stipe 47–152 mm long, with simple eglandular hairs 1.2–3.2 mm long. Lamina narrowly lanceolate, 120–300+ mm long, 20–50+ mm wide, bluntly acute at apex, long-attenuate to broadly cuneate at base, deeply pinnately divided to a distance 0.4–1.3 mm from the midrib, pinnae 30–48+ pairs, lowest pair of pinnae occasionally reduced to auricles, usually 4–5 mm long, longest pinnae 11–26 mm long, 4–8 mm wide, narrowly oblong-triangular to narrowly oblong, obtuse to acute at apex; with simple eglandular hairs 0.6–2.3 mm on both surfaces and on margin; lateral vein branchlets 1–2-forked, free or occasionally anastomosing at apex within and between branch pairs. Sori on 29–35+ pairs of pinnae, ± throughout lamina, 6–10 in each row on longest pinnae, apical $^1/_5$ to ± throughout pinna, midway between pinna mid-vein and margin. Fig. 5/1–5/4.

TANZANIA. Morogoro District: Uluguru Mts, Bondwa [Bondua], 24 Nov. 1932, *Schlieben* 3017! & Uluguru Mts, 29 Aug. 1970, *Pócs* 6231/K! & Uluguru Mts, summit of Bondwa, 26 Sept. 1970, *Faden et al* 70/647!
DISTR. **T** 6; endemic to Uluguru Mts
HAB. Pendulous epiphyte in high altitude mist forest, growing with bryophytes; 2050–2100 m

SYN. [*Xiphopteris villosissima* sensu Iversen in Dansk Bot. Arkiv 7: 151 t. 6 (1932), non (Hook.) Alston]
    Polypodium villosissimum Hook. var. *majus* Reimers in N.B.G.B. 11: 937 (1933)
    [*Xiphopteris villosissima* sensu Schelpe in Contr. Bolus Herb. 1: 8 (1969) pro parte, *non* (Hook.) Alston].]
    [*Zygophlebia villosissima* sensu R.J. Johns, Pterid. Trop. E. Afr. 49 (1991), *non* (Hook.) L.E.Bishop]

NOTES. *Zygophlebia major* is sympatric with *Z. devoluta* in Tanzania, but occurs at higher altitudes.

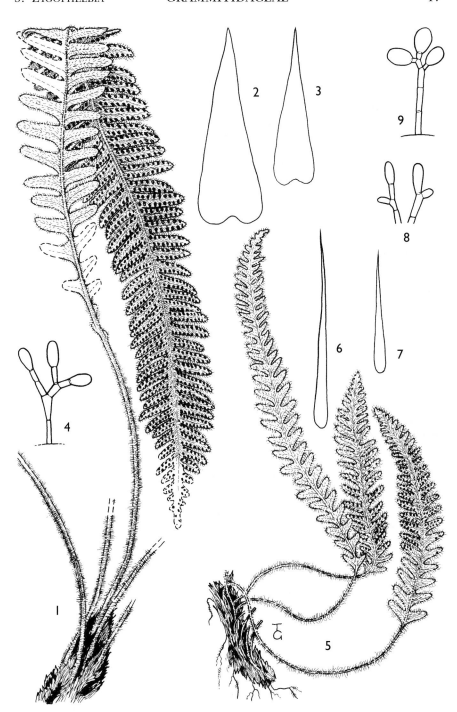

Fig. 5. *ZYGOPHLEBIA MAJOR* — **1**, habit, × 0.7; **2–3**, rhizome scales, × 12; **4**, receptacular paraphysis, × 80. *ZYGOPHLEBIA DEVOLUTA* — **5**, habit, × 0.7; **6–7**, rhizome scales, × 12; **8**, receptacular paraphysis with two glands, × 80; **9**, receptacular paraphysis with three glands, × 80. 1 from *Schlieben* 3018, 2 from *Faden* 70/647, 3 from *Pocs* 6231/K; 4 from *Stolz* 883, 5–8 from *Faden* 74/392; 9 from *Schippers* T1583. Drawn by Tim Galloway.

2. **Zygophlebia devoluta** (*Baker*) *Parris* in K.B. 57, 2: 432 (2002). Type: Madagascar, *Pool* s.n. (K!, holo.)

Rhizome short-creeping, not branched. Stipes 0.1–1.1 mm apart in each row; scales narrowly lanceolate to linear-lanceolate, 2.2–6.7 mm long, 0.3–0.5 mm wide. Stipe 12–93 mm long, with simple eglandular hairs 0.9–3.4 mm long. Lamina narrowly elliptic to narrowly lanceolate in outline, 62–169 mm long, 15–27 mm wide, apex bluntly acute, ± truncate to attenuate at base, pinnate or deeply pinnately divided to a distance 0.1–0.6 mm from the midrib, pinnae 14–47 pairs and 1–4 pairs of apical lobes, lowest pair of pinnae sometimes reduced to auricles, usually 2 mm or more long, longest pinnae 6–15 mm long, 2–5 mm wide, narrowly oblong to narrowly oblanceolate, bluntly acute to obtuse at apex; with simple eglandular hairs 0.8–3.7 mm on both surfaces of lamina and on margin, and translucent 1–2-forked catenate glandular hairs 0.1–0.2 mm as receptacular paraphyses; lateral vein branchlets free, or branchlets of the same vein branch sometimes anastomosing at apex near margin. Sori on 12–37 pairs of pinnae, in apical $^1/_3$ to ± throughout lamina including basal pair of pinnae, 1–9 in each row on longest pinnae, apical $^1/_5$ to ± throughout pinnae, midway between pinna mid–vein and margin. Fig. 5/5–5/9.

TANZANIA. Lushoto District: W Usambaras, Shagayu Forest Reserve, 20 Oct. 1986, *Schippers* T 1583!; Morogoro District: Uluguru Mts, road from Morningside to Bondwa summit, 3 April 1974, *Faden* 74/392!; Rungwe District: Kyimbila, Lukeke, 23 Sept. 1911, *Stolz* 883!
DISTR. **T** 3, 6, 7; Madagascar
HAB. Pendulous epiphyte sometimes high up, in montane forest with *Parinari excelsa, Newtonia buchananii, Ocotea usambarensis, Macaranga kilimandscharica*, sometimes growing with bryophytes, *Mecodium, Elaphoglossum*; 1600–1900 m

SYN. *Polypodium devolutum* Baker in J.L.S. 15: 419 (1876)
    Ctenopteris devoluta (Baker) Tardieu in Notul. Syst. (Paris) 15(4): 445 (1959)
    [*Xiphopteris villosissima* sensu Schelpe in Contr. Bolus Herb. 1: 8 (1969) pro parte, *non* (Hook.) Alston]
    [*X. flabelliformis* sensu Schippers in Fern Gaz. 14(5): 192 (1993), *non* (Poir) Schelpe]

NOTES. A small sterile plant from the West Usambaras (NW on crest of Kisimba above Mazumbai, April 1926, *Peter* 16491 (B)) probably belongs here. It has identical rhizome scales, but rather sparse laminar indumentum which may reflect the aged nature of the fronds.
    Zygophlebia devoluta is the East African and Madagascan vicariant of *Z. villosissima*, which differs in broader rachis wing and shorter proportionately broader pinnae.
    Stolz 883 is more densely hairy than the other two Tanzanian collections, while the type from Madagascar has slightly shorter lamina hairs than Tanzanian material.
    The spores of *Faden* 74/392 (K) are very sparse and largely aborted; the young sporangia appear to be empty in K and MO collections. Likewise the spores of *Stolz* 883 (B, K) are largely misshapen.

6. CERADENIA

L.E.Bishop in Amer. Fern J. 78(1): 2–4 (1988)

Usually epiphytic, rarely terrestrial or lithophytic; rhizomes short-creeping, dorsiventral, stipes in two rows, articulated to rhizome, phyllopodia present; rhizomes with brown non-clathrate scales that are usually glandular at least on margin. Fronds pinnately divided to pinnate; with red-brown simple eglandular hairs, catenate branched hairs often with glandular apex to branches, and simple glandular and catenate simple glandular hairs; veins pinnately branched in pinnae, branchlets often forked, sometimes anastomosing; hydathodes absent on vein endings on adaxial surface of lamina. Sori on surface of lamina or slightly sunken in broad shallow depressions, ± circular to broadly elliptic in outline, in two rows per pinna, one each side of pinna mid-vein, with simple or branched glandular receptacular paraphyses, the glands always white when young; sporangia glabrous.

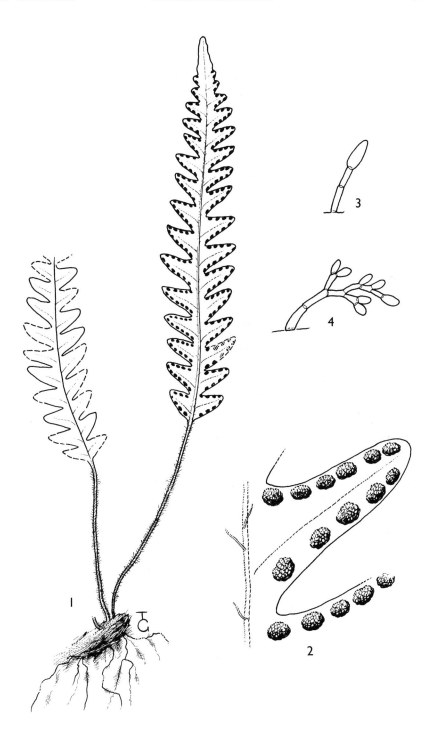

Fig. 6. *CERADENIA SECHELLARUM* — **1**, habit, × 1; **2**, sori on pinnae, × 6; **3**, simple receptacular paraphyses, × 110; **4**, branched receptacular paraphyses, × 110. 1–4 from *Hall* s.n. Drawn by Tim Galloway.

A genus of about 60 species, occurring in tropical montane areas of the New World and Africa including Madagascar.

Ceradenia sechellarum (*Baker*) *Parris* in K.B. 57, 2: 425 (2002). Type: Seychelles, Mahé, *Horne* 194 (K!, lecto. chosen by Morton in Amer. Fern J. 60(3): 124 (1970))

Rhizome short-creeping, not branched. Stipes ± 1 mm apart in each row, rhizome with dense white glandular indumentum amongst scale and rhizoid bases; scales narrowly to linear-lanceolate, with white glandular hairs ± 0.1 mm long on adaxial surface. Stipe 42–58 mm long, with simple eglandular hairs 0.2–2.4 mm long and simple glandular hairs ± 0.1 mm long, sometimes with white exudate. Lamina narrowly lanceolate, 56–104 mm long, 12–20 mm wide, obtuse at apex, broadly cuneate at base, deeply pinnately divided to a distance 0.7–1.3 mm from the midrib, pinnae 13–15 pairs and 2–5 pairs of apical lobes, lowest pairs of pinnae not reduced to auricles, more than $\frac{1}{2}$ as long as longest pinnae, longest pinnae narrowly triangular, 4–14 mm long, 3–8 mm wide, obtuse to acute at apex; with simple eglandular hairs 0.3–2 mm long on abaxial surface of rachis and on margin; lateral veins free. Sori with white gland exudate evident when young, throughout lamina, 6 in each row on longest pinnae, in apical $\frac{2}{3}$ of pinnae to ± throughout, much nearer to margin than pinna mid-vein. Fig. 6.

TANZANIA. Lushoto District: Shagayu Forest, Mt Kwashemhambo, 20 Oct. 1986, *Schippers* T1583b!; Morogoro District: N Uluguru Forest Reserve, Lupanga Peak, E side, 1981, *Hall* s.n.!
DISTR. **T** 3, 6, 7? (see Note); Seychelles
HAB. In moist forest with *Balthasaria, Symphonia, Allanblackia uluguruensis*; 1500–2000 m

SYN. *Polypodium sechellarum* Baker in Hook. & Baker, Syn. Fil. ed. 2: 508 (1874)
    P. albobrunneum Baker, Fl. Maurit., 505 (1877). Type: Seychelles, Mahé, *Horne* 682 (K!, lecto., chosen by Parris in 2002)
    Ctenopteris albobrunnea (Baker) Tardieu in Not. Syst. (Paris) 15(4): 445 (1959)
    Xiphopteris albobrunnea (Baker) Schelpe in Contr. Bolus Herb. 1: 7 (1969); R.J. Johns, Pterid. Trop. E. Afr.: 48 (1991); Schippers in Fern Gaz. 14(5): 192 (1993)
    Grammitis sechellarum (Baker) C.V.Morton in Amer. Fern J. 60(3): 124 (1970)

NOTE. *Ceradenia sechellarum* is related to *C. comorensis* (Baker) Parris which has sori midway between pinna mid-vein and margin, while the former has sori much closer to the margin than to the pinna mid-vein.
    Tanzanian material differs from Seychelles collections in lacking longer simple eglandular hairs on the abaxial surface of the pinna mid-vein, and in having simple glandular hairs on the adaxial surface of the rhizome scales rather than on the margin.
    Schippers in Fern Gaz. 14(5): 192 (1993) mentions what is probably this taxon (as *X. albobrunnea*) for Iringa District: Udzungwa Mts.

INDEX TO GRAMMITIDACEAE

No new names validated in this part

21

GEOGRAPHICAL DIVISIONS OF THE FLORA

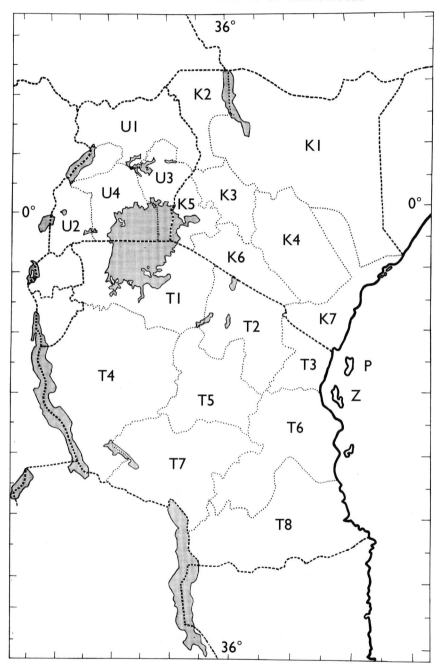

PLANTS PEOPLE
POSSIBILITIES

First published in 2005 by
Royal Botanic Gardens, Kew
Richmond, Surrey, TW9 3AB, UK
www.kew.org

ISBN 1 84246 107 9

Design by Media Resources, typesetting and page layout by Margaret Newman,
Information Services Department,
Royal Botanic Gardens, Kew.

Printed by Cromwell Press Ltd.

For information or to purchase all Kew titles please visit
www.kewbooks.com or email publishing@kew.org